Title Page

The Selenium–Mercury Paradigm

Protection, Pregnancy, and the Chemistry We Misunderstood

Joe Cozart

GMJoe™ Consulting

ISBN: 978-1-970547-13-9

Copyright Page

This work is intended for educational and informational purposes. It does not constitute medical advice, diagnosis, or treatment. Clinical decisions should be made by qualified healthcare professionals based on individual patient circumstances.

Printed in the United States of America.

Foreword

The questions explored in this volume sit at the intersection of chemistry, development, and responsibility. They are not new questions, but they are questions that have been framed narrowly for too long. When guidance persists without fully accounting for the systems it governs, outcomes become harder to explain and harder to improve.

This work does not argue for novelty. It argues for completeness. By restoring selenium to its proper place in the discussion of mercury, diet, and fetal development, it offers a frame that resolves long-standing contradictions without resorting to alarm or simplification. The result is not a reversal of caution, but a clarification of protection.

The value of this contribution lies in its restraint. It does not instruct. It does not moralize. It simply aligns guidance with chemistry and development as they actually operate. That alignment is long overdue.

Author's Note

Author's Note

This book emerged from a sustained examination of how protective systems are often overlooked in favor of more visible threats. Selenium's role in human biology—particularly during pregnancy—has been well documented, yet persistently under-integrated into public and clinical discourse. The pages that follow are an attempt to restore that integration.

No effort has been made to simplify the underlying chemistry beyond what clarity requires. The intended audience includes clinicians, researchers, policy professionals, and others engaged in maternal and developmental health. Where the argument is quiet, it is because the evidence does not require amplification. Where it is firm, it is because the consequences of misunderstanding are real.

The Selenium Mercury Paradigm

Protection rarely announces itself. The systems that matter most operate below the threshold of attention, preserving function long before failure becomes visible. They do not correct damage once it has occurred; they prevent damage from taking shape at all. Selenium belongs to this class of systems. It is not a supplement in the conventional sense, nor a therapeutic intervention applied after harm is identified. It is an infrastructural element of biological defense, embedded in enzymatic processes that sustain neural integrity, regulate oxidative stress, and preserve developmental continuity. Its significance lies not in dramatic effect, but in quiet sufficiency—present before damage is named, before risk is framed, and before fear organizes itself into guidance.

Modern nutritional discourse rarely speaks in this register. We prefer interventions that arrive late and loudly, that announce their necessity through crisis and correction. We are more comfortable reacting than sustaining, more adept at managing failure than preventing it. As a result, the systems that work best often receive the least attention, while the substances most associated with harm dominate the narrative long after their explanatory power has been exhausted. Selenium's obscurity is not accidental. It reflects a broader discomfort with forms of protection that do not lend themselves to alarm or simplification.

For decades, public conversation around seafood has been shaped almost entirely by the presence of mercury.

The story became familiar enough to harden into reflex: mercury is toxic, fish contains mercury, therefore fish—especially for pregnant women—must be approached with caution or avoided altogether. This framing had the advantage of simplicity, but it came at a cost. By isolating mercury as the sole variable of concern, it obscured the biochemical context in which mercury actually operates. It replaced relational chemistry with a single-agent narrative, and in doing so, displaced a far more consequential element from the conversation entirely.

Selenium does not negate mercury by magic or metaphor. It does so by chemistry. Within the body, selenium is required for the synthesis of a family of enzymes responsible for protecting tissues—especially neural tissue—from oxidative damage. These enzymes are not ancillary; they are essential. They operate continuously, preserving cellular integrity in environments that would otherwise degrade rapidly under metabolic stress. Mercury's primary danger lies not in indiscriminate toxicity, but in its affinity for selenium. When mercury binds to selenium, it disrupts the body's ability to produce these protective enzymes.

The harm historically attributed to mercury exposure is, in biochemical terms, a consequence of selenium deprivation.

This distinction matters, because it reframes the entire problem. Mercury does not function as a free agent in biological systems. Its capacity to cause damage depends on whether it encounters sufficient selenium to neutralize its interference. Where selenium is abundant, mercury's threat collapses. Where selenium is depleted, vulnerability emerges. The body does not experience toxicity in the abstract; it experiences imbalance. And imbalance is not measured by the presence of a single element, but by the relationship between elements competing for the same biochemical role.

Fish, particularly ocean fish, occupy a crucial position in this dynamic. Far from being mere carriers of mercury, most commercially consumed species contain selenium in quantities that exceed their mercury content by a substantial margin. This is not a marginal effect or a statistical curiosity. It is a consistent pattern observed across species, geographies, and ecosystems. The fish most commonly eaten by humans—salmon, tuna, sardines, cod, mackerel—are not chemically neutral. They are selenium-positive systems. When consumed, they deliver not only omega-3 fatty acids and high-quality protein, but also the very element required to maintain the body's protective capacity in the presence of trace mercury.

This reality stands in sharp contrast to the way seafood risk has been communicated. Public advisories have tended to treat mercury concentration as a standalone metric, detached from the biochemical environment in which it will be processed. Charts list micrograms without context. Warnings invoke thresholds without explaining mechanism. The result is a narrative that appears cautious while remaining incomplete. It is not that the concern was fabricated; it is that it was framed without regard for the system in which the concern would be resolved.

Protection, in biological terms, is rarely about elimination. It is about sufficiency. The body does not require the absence of challenge to function; it requires the resources to manage challenge without loss of integrity. Selenium provides that resource. It ensures that the enzymes responsible for defending neural tissue remain active, that oxidative stress is neutralized before it cascades into damage, and that essential fatty acids can perform their structural role in the brain without degradation. In this context, fish consumption does not represent a risk to be tolerated. It represents a protective mechanism actively engaged.

The failure to recognize this has had consequences beyond miscommunication. When selenium is excluded from the framework, mercury assumes a prominence it does not deserve. Fear becomes portable, detached from evidence, and resistant to correction. Over time, the narrative ossifies. What began as provisional caution hardens into doctrine, and doctrine persists long after the data that justified it has been superseded. Meanwhile, the protective systems that were always present remain unacknowledged, doing their work without recognition.

Understanding selenium as protection rather than supplement changes the posture of the entire discussion. It shifts the focus from avoidance to adequacy, from restriction to provision. It also clarifies why populations with high seafood consumption have not exhibited the cognitive harm that mercury-centered models predicted. The explanation is not cultural anomaly or genetic exception. It is biochemical sufficiency. These populations consumed fish that delivered selenium alongside omega-3 fatty acids, maintaining the integrity of the very systems mercury would otherwise disrupt.

This reframing does not minimize mercury. It places it in proportion. Mercury remains relevant insofar as it competes with selenium for biological binding. But it is no longer the axis around which dietary guidance must revolve. The axis shifts instead to the maintenance of protective capacity—ensuring that selenium intake is sufficient to sustain enzymatic function, particularly during periods of heightened vulnerability.

No period is more vulnerable than pregnancy. During fetal development, the demand for selenium increases dramatically, while reserves are effectively nonexistent. The developing brain cannot store selenium for later use; it depends entirely on continuous transfer from the mother. Any disruption to that transfer—whether through deficiency, displacement, or misinformed dietary restriction—has immediate and irreversible consequences. In this context, the absence of selenium is not a neutral condition. It is a deficit with structural implications.

By approaching the issue through the lens of protection rather than toxicity, the contours of responsibility become clearer. The question is no longer how to eliminate every trace of mercury from the diet—a biological impossibility—but how to ensure that the systems designed to manage trace exposure remain intact. Fish, when understood properly, emerge not as a hazard to be managed, but as one of the most efficient delivery systems for that protection.

This is the paradigm that has been missing. Not because it was hidden, but because it did not align with the incentives of simplification. Selenium does not lend itself to alarmist headlines. It does not generate fear. It does not demand immediate behavioral reversal. It simply operates, quietly maintaining balance where imbalance would otherwise take hold. To recognize its role requires a willingness to think relationally, to accept that risk and protection coexist within the same system, and that eliminating one variable without regard for its counterpart can produce outcomes worse than the original threat.

The chemistry was never complicated. What failed was the frame.

Mercury persists in public imagination as a singular villain because it is easy to isolate and difficult to contextualize. It presents cleanly as a threat, lends itself to numerical thresholds, and invites the reassuring logic of avoidance. But biology does not operate through isolated actors. It operates through competition, affinity, and displacement. Mercury's significance, when understood correctly, is not that it is uniquely destructive, but that it interferes with a system already designed to protect against disruption. It is not an agent of chaos in the abstract; it is a relational disruptor whose effects depend entirely on what it encounters.

The mistake embedded in single-variable toxicology is not simply that it overstates danger. It is that it misidentifies mechanism. When mercury enters the body, it does not roam freely attacking tissues at random. Its primary biochemical behavior is to bind.

Specifically, it binds to selenium with remarkable affinity, forming complexes that render selenium unavailable for its biological role. This binding does not create toxicity by addition; it creates vulnerability by subtraction. The body's capacity to defend itself is compromised not because mercury introduces a new harm, but because it removes a preexisting protection.

This distinction reshapes how exposure should be interpreted. In a system where selenium is abundant, mercury's binding behavior is largely neutralized. The protective enzymes that depend on selenium continue to be synthesized, oxidative stress remains controlled, and neural tissues retain their structural integrity. In a system where selenium is scarce or displaced, the same level of mercury exposure produces a markedly different outcome. The difference is not dose alone. It is context.

Single-variable models struggle with context because context complicates guidance. It is far easier to issue a warning based on concentration than to explain relational sufficiency. Charts listing mercury levels in isolation offer an illusion of precision while omitting the chemistry that determines biological effect.

In this framework, fish species are ranked by risk without regard for their selenium content, and consumers are instructed to limit intake based on an incomplete metric. The result is guidance that appears cautious but is fundamentally misaligned with how harm actually manifests.

The persistence of this model has less to do with evidence than with institutional inertia. Toxicology developed in an era when isolating variables was necessary for measurement. What began as a methodological convenience hardened into interpretive habit. Over time, the habit became doctrine. Mercury was elevated from disruptor to determinant, and selenium was relegated to obscurity—not because its role was disproven, but because it complicated the narrative. Once the frame was set, new data struggled to dislodge it, even when that data explained inconsistencies the old model could not.

Those inconsistencies have always been visible. Populations with high fish consumption have not exhibited the neurological deficits predicted by mercury-centric models. Instead of prompting a reevaluation of assumptions, these outcomes were often treated as anomalies—exceptions to be explained away rather than signals that the framework itself was incomplete. The relational chemistry was present, observable, and replicable, yet remained peripheral to policy and public guidance.

Reframing mercury as a relational disruptor resolves these contradictions without special pleading. It explains why mercury exposure does not correlate linearly with harm, why selenium-rich diets confer resilience, and why avoidance strategies that reduce selenium intake can paradoxically increase risk. It also clarifies why attempts to refine mercury thresholds have produced diminishing returns. No threshold can account for a relationship it refuses to model.

This reframing also alters how risk should be communicated. When mercury is treated as the central variable, the logical response is restriction. When it is understood as a disruptor whose impact depends on selenium availability, the logical response becomes provision. Ensuring adequate selenium intake is not a secondary consideration; it is the primary determinant of outcome. Fish consumption, in this context, is not a gamble between benefit and harm. It is a delivery mechanism that sustains the very protection mercury would otherwise compromise.

The language of toxicity obscures this reality because it implies inevitability. Toxins are assumed to cause damage by their mere presence. Relational disruptors do not. They require conditions. They exploit deficiencies.

They reveal vulnerabilities that already exist. Mercury's danger, therefore, is not intrinsic; it is conditional. And conditions can be managed.

Understanding this conditionality is essential when evaluating dietary guidance that emphasizes avoidance without substitution. Removing fish from the diet does not simply remove mercury; it removes selenium, omega-3 fatty acids, and a suite of micronutrients integral to neural health. The net effect is not neutral. It shifts the system toward deficiency, increasing susceptibility to the very disruptions the guidance sought to prevent. The irony is rarely acknowledged, but it is consequential.

The reluctance to adopt this relational model reflects a broader discomfort with explanations that resist simplification. Single-variable toxicology offers clarity at the expense of accuracy. Relational chemistry offers accuracy at the expense of easy messaging. Institutions tasked with protecting public health often default to the former, not out of malice, but out of structural preference. Fear travels faster than nuance. Avoidance is easier to prescribe than sufficiency. And once a warning has been issued, retracting or revising it carries reputational risk.

Yet the cost of maintaining an incomplete model is not abstract. It accumulates quietly, particularly in populations for whom nutritional adequacy is already precarious. When mercury is framed as the dominant concern, selenium deficiency becomes invisible. When deficiency becomes invisible, its consequences are misattributed or ignored. The body's protective systems fail not because they were overwhelmed, but because they were never adequately supported.

This is not an argument for dismissing mercury. It is an argument for understanding it correctly. Mercury matters because it competes. It disrupts by binding. It exerts influence by displacing an element essential to defense. Treating it as an isolated toxin obscures the leverage point where intervention is most effective. Treating it as a relational disruptor reveals that leverage point clearly and consistently.

Once this perspective is adopted, the apparent complexity of the issue dissolves. The chemistry is straightforward. Selenium sufficiency preserves enzymatic protection. Mercury interference becomes relevant only when that sufficiency is compromised. Fish consumption, rather than representing a tradeoff between benefit and risk, emerges as a stabilizing factor within this system. The problem was never fish. The problem was the frame through which fish were evaluated.

This reframing also exposes the limitations of incremental policy adjustments that leave the underlying model intact. Refining mercury advisories without integrating selenium does not resolve the core error; it perpetuates it. As long as mercury is treated as the primary variable, guidance will continue to misalign with biological reality. The solution is not better charts or more granular thresholds. It is a shift in paradigm—from isolation to relationship, from fear to function.

Relational biology demands a different kind of literacy. It requires acknowledging that elements do not act alone, that harm and protection are often two sides of the same interaction, and that eliminating one variable without sustaining its counterpart can destabilize the entire system. Mercury's role, properly understood, reinforces this lesson rather than contradicting it. It teaches that vulnerability is created not by presence alone, but by imbalance.

With this understanding in place, the path forward becomes clearer. The question is no longer how low mercury exposure must be driven to achieve safety. It is how consistently selenium sufficiency can be maintained to preserve protection. That shift in question marks the true boundary between outdated toxicology and a model aligned with how the body actually works.

Pregnancy introduces a condition unlike any other in human biology: a period in which demand accelerates while reserves do not exist. The developing fetus does not draw from a stockpile, nor does it possess redundancy. Every requirement must be met in real time, through continuous transfer, across a placental boundary that is selective but not forgiving. In this environment, adequacy is not optional, and delay is not recoverable. What is not provided when needed is not compensated for later. It is simply absent, and the consequences of that absence are carried forward.

Selenium occupies a particularly unforgiving position within this economy. Unlike some nutrients that can be stored in meaningful quantities, selenium functions primarily through immediate incorporation into enzymes that must be synthesized continuously. The fetal brain, undergoing rapid growth and differentiation, relies on these enzymes to control oxidative stress, preserve membrane integrity, and protect long-chain fatty acids essential to neural architecture. There is no surplus capacity. Selenium delivered today supports protection today. Selenium not delivered cannot be reclaimed tomorrow.

This temporal rigidity is what makes pregnancy a one-way biochemical window. The body does not pause development while awaiting sufficiency. Neurons form, migrate, and connect according to a schedule dictated by biology, not by dietary correction. When selenium availability is disrupted during this process—whether through deficiency, displacement, or misguided avoidance—the resulting vulnerability is embedded in structure, not merely function. Once that window closes, the opportunity to intervene has passed.

Public health guidance has struggled to accommodate this reality because it resists simplification. Pregnancy advice often treats nutrients as interchangeable inputs and risks as cumulative exposures. In this framework, avoiding a perceived toxin appears prudent, even if that avoidance simultaneously eliminates an essential protective element. The logic feels conservative, but it fails to account for asymmetry. A missed exposure can sometimes be tolerated. A missed protection during development cannot.

Mercury-centered advisories illustrate this failure with particular clarity. When pregnant women are advised to limit or avoid fish consumption without explicit attention to selenium sufficiency, the guidance inadvertently prioritizes hypothetical exposure over actual protection. The fetus does not experience benefit from avoidance alone. It experiences benefit only from adequacy—of selenium, of essential fatty acids, of the enzymatic systems that preserve neural integrity during rapid growth.

The placenta does not buffer this imbalance. Selenium transfer from mother to fetus is active and preferential, reflecting its importance, but it is not immune to competition. When mercury binds selenium in the maternal system, the pool available for transfer is reduced. When dietary intake is restricted, the pool contracts further. The combined effect can be subtle and easily overlooked, yet its impact is magnified by the absence of fetal reserves. What the fetus receives is what exists. There is no margin for error.

This reality exposes a central flaw in risk models that treat pregnancy as merely a higher-sensitivity version of adult physiology. It is not. Pregnancy is a distinct biochemical state with different priorities, different vulnerabilities, and different failure modes. Strategies that might be neutral or even beneficial for non-pregnant adults can be actively harmful when applied during gestation. Avoidance without substitution is one such strategy. It assumes that removing risk leaves a neutral baseline. In pregnancy, removal without replacement leaves a deficit.

The consequences of selenium insufficiency during gestation are not always dramatic or immediately visible. They do not announce themselves as acute toxicity. Instead, they manifest as subtle compromises in neural resilience—changes in how the brain responds to stress, how it processes information, how it maintains structural integrity over time. These outcomes are difficult to trace back to a single cause, which is precisely why they persist uncorrected. They fall into the category of damage that does not provoke alarm because it lacks spectacle.

This invisibility has allowed a persistent misunderstanding to endure: that risk during pregnancy can be managed primarily through restriction. In reality, risk is managed through balance. The fetus does not benefit from a reduced intake of protective nutrients simply because a potential disruptor has also been reduced. It benefits only when protection is sustained at the level required by developmental demand.

Fish consumption during pregnancy exemplifies this principle. Fish provide selenium in a form and proportion that aligns with biological need, alongside omega-3 fatty acids critical for brain development. When consumed, they support the enzymatic systems that guard against oxidative stress at precisely the moment those systems are under greatest strain. Removing fish from the maternal diet removes this support, replacing a biologically coherent delivery system with a void that must somehow be filled by alternative means. Often, it is not.

The failure to articulate this tradeoff explicitly has left many pregnant women navigating guidance that emphasizes caution without clarity. Lists of "safe" and "unsafe" species obscure the underlying chemistry. Thresholds are presented without context. The implicit message is that less exposure equals more safety. But in the absence of selenium sufficiency, less exposure can mean less protection, and less protection during fetal development carries consequences that no later intervention can reverse.

This is not a call for recklessness. It is a call for precision. The fetal brain does not require heroic levels of any single nutrient. It requires continuity. Selenium must be present consistently, in adequate amounts, throughout gestation. Disruptions—whether through dietary deficiency, competitive binding, or institutional guidance that undervalues its role—translate directly into vulnerability. The window does not reopen.

Understanding pregnancy as a one-way biochemical economy clarifies why correcting course after birth is insufficient. Postnatal nutrition supports growth and function, but it cannot reconstruct pathways that were never properly protected during formation. The architecture of the brain reflects the conditions under which it developed. Selenium sufficiency during gestation is part of that architecture, not an accessory to it.

This perspective also reframes responsibility. The burden does not rest on individual women to decipher complex biochemistry in the face of ambiguous advice. It rests on the systems that produce that advice to model the chemistry accurately. When guidance focuses narrowly on avoidance, it abdicates that responsibility. When it integrates protection, it fulfills it.

Pregnancy tolerates neither delay nor approximation. It demands that sufficiency be achieved when needed, not recognized in retrospect. Selenium's role in this demand is not peripheral. It is foundational. To treat it as optional, secondary, or interchangeable is to misunderstand the nature of development itself.

The cost of that misunderstanding is not measured in immediate crises, but in lost potential—quietly, cumulatively, and permanently. In a system where protection must precede harm, pregnancy is the clearest example of why timing matters more than intention. What is not provided in time is not merely delayed. It is gone.

Institutions rarely fail by choosing the wrong goal. They fail by committing too early to a frame and then optimizing relentlessly within it. Once a frame is established, evidence is filtered through it, language conforms to it, and authority becomes invested in maintaining its coherence. The longer the frame persists, the more difficult it becomes to revise—not because the data are unclear, but because revision implies that past guidance was incomplete. In public health, incompleteness is often treated as error, and error as liability. The result is not stagnation, but drift: a slow movement away from biological reality toward narrative stability.

The mercury-centered model of dietary risk emerged from a legitimate concern. Toxicology, as a discipline, required isolation to measure effect. Mercury was measurable, traceable, and plausibly linked to neurological harm. Selenium, by contrast, operated quietly, relationally, and without obvious thresholds. It did not announce itself through acute toxicity, nor did it lend itself to simple warnings. In the early stages of policy formation, this asymmetry mattered. What could be measured and communicated with confidence took precedence over what complicated the message.

Over time, provisional caution hardened into institutional memory. Advisories were written, charts were published, and risk categories were established. Each iteration reinforced the same underlying assumption: mercury concentration alone defined danger. The absence of selenium from this framework was not the result of deliberate exclusion. It was the byproduct of a system that privileges clarity over completeness. Once the guidance existed, it became self-reinforcing. Updating it required not only new data, but a willingness to revisit the assumptions that made the original guidance legible.

This reluctance is not unique to nutrition. It reflects a broader pattern in how institutions manage uncertainty. When faced with complex, relational systems, there is a tendency to reduce them to single variables that can be monitored and controlled. The reduction creates the appearance of rigor, even as it strips away explanatory power. When contradictions emerge—when outcomes do not match predictions—the system responds by refining the variable rather than questioning the frame. Thresholds are adjusted. Categories are subdivided. The core assumption remains untouched.

In the case of seafood guidance, this approach produced increasingly detailed lists of acceptable and unacceptable fish, differentiated by species, size, and origin. What it did not produce was a model that accounted for the chemistry governing harm and protection. Selenium remained absent not because its role was unknown, but because integrating it would require a fundamental shift from avoidance-based messaging to sufficiency-based guidance. That shift would complicate communication, redistribute responsibility, and challenge the moral clarity of warnings that had already been internalized by the public.

Fear, once established, acquires its own momentum. It is easier to sustain than nuance, easier to defend than correction. Advisories framed around avoidance create a clear moral signal: caution equals care. Revising those advisories to include protection risks being misinterpreted as relaxation or negligence, even when the revision is scientifically justified. Institutions tasked with safeguarding health are acutely sensitive to this perception. The safest position, reputationally, is often to maintain the warning and refine it incrementally rather than to reframe it entirely.

This dynamic has consequences. By centering mercury and excluding selenium, guidance inadvertently trains both clinicians and patients to think in terms of subtraction rather than balance. Risk becomes something to eliminate rather than something to manage through adequacy. In this environment, nutrients that do not announce themselves through deficiency syndromes are undervalued. Selenium deficiency does not present with a single, dramatic symptom that demands immediate correction. Its effects are distributed, subtle, and often delayed. They are therefore easy to overlook and difficult to attribute.

The result is a persistent mismatch between policy and physiology. The body operates relationally, integrating competing elements to preserve function. Guidance operates atomistically, isolating variables to preserve clarity. When these two logics diverge, the cost is borne not by the system that issued the guidance, but by the populations that live under it. In pregnancy, where margins are narrow and timing is unforgiving, that cost becomes particularly consequential.

Institutional narratives also shape clinical practice. When guidelines emphasize mercury avoidance without equal emphasis on selenium sufficiency, clinicians are left with an incomplete framework. Even those aware of the broader chemistry may hesitate to depart from official recommendations, particularly in environments where deviation carries professional risk. The guidance becomes a ceiling rather than a baseline—something to adhere to rather than interrogate. Over time, the absence of selenium from the narrative becomes normalized, and its importance fades from clinical consideration.

This normalization is reinforced by the way knowledge is transmitted. Medical education often compartmentalizes nutrition, toxicology, and obstetrics into separate domains. Selenium's role sits at the intersection of all three, making it vulnerable to omission. Mercury, by contrast, fits neatly into toxicology modules and risk communication frameworks. The institutional structure itself favors one variable over the other, not because it is more important, but because it is easier to place.

None of this requires bad faith to persist. In fact, it persists precisely because it is well-intentioned. The system believes it is protecting the public by erring on the side of caution. What it fails to recognize is that caution, when defined narrowly, can produce its own form of harm. Avoidance without adequacy is not neutral. It reshapes diets, alters nutrient profiles, and introduces deficiencies that the guidance never intended to create.

Correcting this drift requires more than adding a footnote or revising a chart. It requires acknowledging that the original frame was incomplete. That acknowledgment is institutionally difficult because it challenges the implicit claim of sufficiency embedded in existing guidance. Yet without it, the system remains locked into a narrative that cannot fully explain outcomes or guide optimal behavior.

The path forward is not to abandon caution, but to refine it. A model that integrates selenium alongside mercury does not weaken protection; it strengthens it. It aligns guidance with the chemistry that governs risk, rather than with the variables that are easiest to communicate. It also restores agency to clinicians, who can then counsel patients based on sufficiency and balance rather than restriction alone.

Institutional change is slow, particularly when it involves revising narratives that have become moralized. But the cost of delay is cumulative. Each generation of guidance that omits selenium reinforces the same misunderstanding, embedding it more deeply in practice and perception. The longer the omission persists, the harder it becomes to correct, not because the evidence is lacking, but because the story has become familiar.

Understanding how this drift occurred is essential, not to assign blame, but to recognize the structural forces that sustain it. Institutions respond to incentives, narratives, and reputational risk as much as to data. When those forces favor simplicity over accuracy, even well-established chemistry can remain sidelined. Bringing selenium back into the frame is therefore not merely a scientific correction. It is an institutional one—a recalibration of how risk, protection, and responsibility are understood.

The failure here was not one of knowledge, but of integration. The chemistry was known. The relationships were observable. What was missing was a willingness to let relational biology displace single-variable comfort. Until that displacement occurs, guidance will continue to orbit a partial truth, leaving protection underarticulated and vulnerability insufficiently addressed.

In systems designed to prevent harm, the most consequential errors are often those that feel safest to maintain. The mercury narrative became one such error—not because it was wrong, but because it was incomplete. Recognizing that incompleteness is the first step toward restoring alignment between institutional guidance and biological reality.

Once protection is restored to its proper place in the frame, diet stops behaving like a list of prohibitions and begins to look like a system of provision. This shift is subtle but decisive. It replaces the logic of subtraction with the logic of sufficiency and allows disparate elements—trace minerals, fatty acids, enzymatic defense—to be understood as cooperating rather than competing. When this cooperation is acknowledged, many of the contradictions that have plagued nutritional guidance dissolve without argument.

Fish occupy a central position in this reassembled model not because they are exceptional, but because they are coherent. They deliver selenium in a form that aligns with biological need, paired with omega-3 fatty acids that require protection from oxidation in order to perform their structural role in neural tissue. This pairing is not incidental. It reflects an evolutionary compatibility between nutrient delivery and enzymatic defense. Selenium-dependent enzymes preserve the integrity of the very lipids that make neuronal membranes flexible, resilient, and capable of rapid signaling. Diet, in this sense, is not a menu of isolated inputs. It is an architecture.

When guidance is framed around avoidance, this architecture is fragmented. Individual components are evaluated in isolation, and the relationships that confer stability are lost. The body is then asked to compensate for what guidance has removed, often without the means to do so. Supplements are proposed as substitutes, thresholds are adjusted, and complexity is pushed downstream onto individuals and clinicians. What is missing from this approach is not data, but synthesis.

Reassembling the model requires acknowledging that nutrients do not act independently and that risk cannot be managed by eliminating a single element without regard for what that element displaces. Selenium sufficiency is not a marginal consideration layered onto otherwise complete advice. It is the condition that allows other benefits to be realized safely. Omega-3 fatty acids, widely recognized for their role in fetal brain development, are particularly vulnerable to oxidative damage in the absence of adequate enzymatic protection. Providing them without ensuring selenium sufficiency is akin to delivering building materials without preserving the scaffolding that holds them in place.

This integrated view also clarifies why attempts to optimize diet through isolated recommendations often produce uneven results. A focus on omega-3 intake without parallel attention to selenium can fail to deliver the expected benefit. Conversely, selenium sufficiency achieved through diet stabilizes multiple pathways simultaneously, reducing the need for micromanagement. The system works when its internal relationships are respected.

For pregnancy, this coherence matters more than at any other time. The developing brain requires not only raw materials, but protection during assembly. Selenium's role in preserving enzymatic defense ensures that growth proceeds without accumulating damage that cannot later be undone. Fish consumption during gestation, when understood through this lens, emerges as a stabilizing practice rather than a calculated risk. It provides what development demands at the moment demand is greatest.

The reluctance to articulate this coherence has left a vacuum in guidance. In the absence of an integrated model, lists and limits proliferate. Species are categorized, servings are capped, and the underlying chemistry remains unspoken. Clinicians are left to interpret charts without a framework, and patients are left to navigate advice that feels both urgent and incomplete. The result is confusion that masquerades as caution.

A forward-facing model does not require instructive language to be effective. It requires clarity. When the relationships are visible, appropriate behavior follows naturally. Diet becomes less about compliance and more about alignment. The question shifts from "What must be avoided?" to "What sustains protection?" That shift alone resolves much of the anxiety surrounding seafood consumption during pregnancy, replacing it with a rationale grounded in biology rather than fear.

This clarity also exposes the limitations of substitution strategies that attempt to replicate whole foods through isolated components. Supplements can address specific deficiencies, but they rarely reproduce the relational balance delivered by coherent dietary sources. Selenium provided through fish arrives in proportion to other nutrients that depend on its presence. This proportionality is difficult to engineer artificially and easy to disrupt through restriction. The body evolved to operate within these proportions, not to assemble them piecemeal.

Reassembling the model does not imply uniformity. It does not suggest that all fish are identical or that context is irrelevant. It suggests that evaluation must occur at the level where harm and protection actually interact. Selenium-to-mercury balance, enzymatic capacity, and developmental timing are the variables that matter. Species lists and numerical thresholds are proxies at best. They cannot substitute for an understanding of the system they attempt to regulate.

The forward-facing model also reframes the role of uncertainty. Instead of treating uncertainty as a reason for maximal restriction, it treats it as a reason to preserve known protective mechanisms. Selenium's role in enzymatic defense is not speculative. Its necessity is well established. Where uncertainty remains—about exposure, about individual variation—the prudent response is to ensure sufficiency, not to erode it. Protection absorbs uncertainty; deficiency amplifies it.

This reframing has implications beyond pregnancy. It challenges a broader tendency in nutrition and public health to privilege simplicity over integration. Systems that rely on relational balance are poorly served by guidance that isolates variables for the sake of communication. The cost of that isolation is not always immediate, but it is cumulative. Over time, populations adapt to guidance that removes coherence from diet, and the burden of adaptation falls on biological systems that were never designed to operate in fragments.

By restoring protection to the center of the frame, diet can once again be discussed as a means of sustaining function rather than managing threat. Fish, selenium, and fatty acids return to their proper relationship. Mercury, understood as a disruptor rather than a determinant, loses its disproportionate influence over guidance. The narrative becomes quieter, but more accurate. It no longer needs to persuade because it explains.

This quiet accuracy is what allows the model to move forward without confrontation. It does not demand that previous guidance be condemned. It simply renders it insufficient. The new frame does not compete with the old one; it subsumes it, accounting for the same concerns while explaining what the older model could not. In doing so, it offers a way out of the impasse between caution and adequacy.

A coherent dietary model does not eliminate risk. It situates risk within a system capable of managing it. Selenium sufficiency is one such capability, essential not because it guarantees safety, but because it preserves resilience. In pregnancy, where resilience must be present from the outset, this preservation is non-negotiable. Development proceeds regardless of guidance. The question is whether protection keeps pace.

By reassembling diet, protection, and development into a single frame, the path forward becomes intelligible. It does not require new warnings or revised lists. It requires a willingness to speak in terms of systems rather than substances, relationships rather than limits. When that willingness is present, the chemistry that was always sufficient becomes visible again, and guidance can finally align with the biology it seeks to serve.

Deficiency rarely announces itself in the way toxicity does. It does not arrive with a singular event, a dramatic threshold, or a clear temporal marker that demands attention. Instead, it alters probabilities. It reshapes outcomes at the margins, nudging systems toward fragility without declaring itself as the cause. Selenium deficiency is exemplary in this regard. Its effects are dispersed across time, distributed among tissues, and easily mistaken for background variation. What it erodes is not function outright, but resilience—the capacity to absorb stress without consequence.

This invisibility has been decisive. In public health, what cannot be readily observed or measured is often deprioritized, regardless of its importance. Selenium does not produce a classic deficiency syndrome that forces recognition. There is no singular symptom that compels intervention. Instead, insufficiency manifests as an increased susceptibility to oxidative stress, a diminished capacity to protect polyunsaturated fatty acids, and a subtle weakening of enzymatic defenses that are most critical during periods of rapid growth or heightened demand. These effects are real, but they are not spectacular. They do not cluster neatly around a diagnosis. They are therefore easy to miss.

When outcomes are subtle, attribution becomes difficult. Developmental differences, cognitive variability, and long-term neurological resilience are influenced by numerous factors. Selenium's role, operating quietly in the background, is easily overshadowed by more conspicuous explanations. Socioeconomic conditions, genetics, education, and environment all present themselves as plausible causes. In this crowded field, a trace element that exerts its influence through protection rather than performance struggles to claim recognition.

The problem is compounded by the way nutritional adequacy is assessed. Many frameworks rely on population averages and overt deficiency markers, neither of which capture the dynamics of protection. Selenium sufficiency is not about meeting a minimal requirement to avoid acute illness. It is about maintaining enzymatic capacity under conditions of stress. A population can appear adequate by conventional measures while still harboring widespread vulnerability at precisely the moments when protection matters most. Pregnancy is one such moment. Rapid neural development amplifies the cost of even modest shortfalls, yet those shortfalls may remain invisible to standard assessments.

This invisibility allows deficiency to masquerade as normal variation. When guidance reduces fish consumption without explicitly replacing selenium, the resulting deficit does not present as a crisis. Instead, it subtly alters developmental trajectories in ways that are difficult to trace back to diet. The absence of immediate consequences reinforces the belief that avoidance was harmless. Over time, this belief becomes entrenched, even as the underlying chemistry continues to operate.

The misattribution extends beyond individuals to institutions. When selenium's protective role is not integrated into the model, observed outcomes are explained through alternative narratives. Mercury retains its status as the presumed culprit, even when evidence fails to support a direct causal link. The persistence of this explanation owes less to its accuracy than to its familiarity. It offers a clear antagonist and a straightforward response. Selenium deficiency, by contrast, offers no villain and no drama. It asks only that protection be maintained consistently—an unremarkable demand that struggles to compete for attention.

This dynamic also shapes research priorities. Studies designed to detect toxicity are more likely to focus on exposure levels than on protective balance. Outcomes are analyzed in terms of harm rather than resilience. When selenium is included, it is often treated as a covariate rather than as a central variable. Its role is acknowledged but not foregrounded. The result is a body of literature that contains the answer without ever assembling it into a coherent paradigm.

Deficiency hides especially well in systems that valorize intervention over maintenance. Medical culture is adept at correcting abnormalities once they are identified. It is less comfortable sustaining conditions that prevent abnormalities from emerging in the first place. Selenium's function aligns with the latter. It preserves the integrity of systems before they fail. In doing so, it deprives clinicians of the very signals—acute dysfunction, dramatic improvement—that typically validate intervention. What works best leaves the least evidence of its necessity.

The consequences of this invisibility are cumulative. When protective systems are undervalued, guidance gravitates toward strategies that feel active but undermine resilience. Avoidance becomes a proxy for care. Supplements are proposed as afterthoughts rather than integrated solutions. The body is asked to tolerate gaps in protection on the assumption that they will be inconsequential. Over time, these gaps compound, particularly in populations where baseline intake is already marginal.

Selenium deficiency also hides because its effects are often delayed. The absence of protection during development may not manifest immediately. It may influence how the brain responds to stress later in life, how it copes with inflammation, or how it maintains structural integrity over decades. These outcomes are rarely linked back to prenatal nutrition, not because the link does not exist, but because the temporal distance obscures it. Cause and effect drift apart, and the opportunity for correction passes unnoticed.

This delay reinforces a dangerous complacency. If no immediate harm is observed, guidance is assumed to be adequate. The system congratulates itself for avoiding visible toxicity while ignoring the quieter erosion of resilience. Selenium's absence from the narrative becomes self-justifying: if deficiency were truly important, the reasoning goes, its effects would be obvious. That reasoning misunderstands the nature of protection. What protection prevents cannot be observed because it does not occur.

The invisibility of deficiency also affects how responsibility is distributed. When harm is diffuse and delayed, it is difficult to assign accountability. Individuals are left to navigate conflicting advice, clinicians defer to guidelines, and institutions maintain existing frames. Selenium falls through the cracks not because anyone intends to neglect it, but because it does not demand attention. It simply waits to be displaced.

Understanding this dynamic is essential to resolving the selenium–mercury paradigm. The persistence of misunderstanding is not primarily a failure of evidence. It is a failure of visibility. Selenium's role has been known, documented, and replicated. What has been missing is a willingness to center protection in a system accustomed to reacting to harm. Until that shift occurs, deficiency will continue to hide in plain sight, shaping outcomes without acknowledgment.

Recognizing hidden deficiency does not require alarmism. It requires a change in emphasis. Instead of asking whether exposure exceeds a threshold, the more relevant question becomes whether protection is sufficient to absorb that exposure. Instead of searching for dramatic endpoints, attention shifts to resilience and capacity. Selenium's contribution becomes legible only when the model allows for such questions.

Once deficiency is made visible, the paradox dissolves. The absence of acute toxicity no longer implies safety, and the presence of trace exposure no longer implies danger. The system is evaluated on its ability to maintain balance rather than eliminate challenge. Selenium, long overlooked because it operates quietly, emerges as central precisely because it sustains that balance.

Deficiency that hides in plain sight is the most costly kind. It invites complacency while eroding protection, and it persists because it does not provoke response. Bringing selenium into focus corrects this blind spot, not by exaggerating its role, but by acknowledging what it has always done. Protection, when it works, leaves no trace. The task is to recognize it anyway.

Population data has a way of revealing what theory alone cannot. When individual mechanisms are difficult to isolate, large patterns often expose the underlying structure by their consistency. In the case of seafood consumption, these patterns have been visible for decades, quietly contradicting the expectations of mercury-centered models while aligning seamlessly with a framework that accounts for protection. The contradiction has not been subtle. It has simply been inconvenient.

Across cultures with high habitual fish intake, neurological outcomes have not followed the trajectory predicted by mercury avoidance logic. Coastal populations, island nations, and fishing communities consume seafood at levels that would be considered excessive under restrictive advisories, yet they do not exhibit the widespread cognitive impairment those advisories imply. Instead, many demonstrate outcomes that are neutral at worst and favorable at best, particularly with respect to neurodevelopment and cardiovascular health. These observations have been repeatedly documented, often accompanied by attempts to explain them away rather than integrate them into the dominant model.

The difficulty lies not in the data, but in the lens through which it is interpreted. When mercury is treated as the primary determinant of risk, populations that consume large amounts of fish should, in theory, display measurable harm. When they do not, the discrepancy is attributed to confounding variables—genetics, lifestyle, or cultural practices—rather than prompting a reassessment of the model itself. These explanations are not wrong in isolation, but they function as deflections. They preserve the centrality of mercury by relegating contradictory outcomes to exception rather than evidence.

A selenium-centered perspective resolves these inconsistencies without strain. Fish consumed by these populations are not merely sources of protein; they are vehicles of selenium sufficiency. The selenium content of commonly consumed species often exceeds their mercury content by a substantial margin, preserving enzymatic defense even in the presence of trace exposure. The relationship is not incidental. It is systemic. Where selenium intake is robust, mercury's disruptive potential is neutralized, and the predicted harm fails to materialize.

This pattern is visible not only across cultures, but across time. Populations with long-standing seafood traditions did not experience a sudden emergence of neurological decline with the introduction of industrial pollutants at the levels typically encountered through diet. Where harm has been observed, it has often coincided with circumstances that disrupt selenium balance—environmental contamination severe enough to overwhelm protective capacity, or dietary shifts that reduce selenium intake while maintaining exposure. These cases do not undermine the paradigm; they confirm it. They demonstrate that protection has limits, but also that those limits are defined by balance, not by exposure alone.

Large-scale epidemiological studies have repeatedly struggled to establish a consistent link between typical dietary mercury exposure and adverse neurodevelopmental outcomes. Findings are mixed, associations are weak, and results often depend on how exposure is measured and which variables are controlled. From a mercury-only perspective, this inconsistency is frustrating. From a relational perspective, it is expected. When selenium sufficiency varies across populations and individuals, mercury's impact cannot be predicted by concentration alone. The signal is diluted not because it is absent, but because it is conditional.

The failure to incorporate selenium into these analyses has led to a peculiar stalemate. Evidence accumulates without resolution, and guidance remains cautious without becoming precise. The absence of clear harm is interpreted as uncertainty rather than as an indication that the model may be incomplete. Meanwhile, the populations most affected by restrictive advice—pregnant women, children, and communities encouraged to reduce seafood intake—bear the nutritional consequences of guidance that errs on the side of fear.

Observed outcomes in pregnancy further reinforce this pattern. Studies examining maternal fish consumption have often found associations with improved neurodevelopmental markers in offspring, even when mercury exposure is present. These findings are typically framed as benefits that outweigh risks, a formulation that implies a tradeoff rather than coherence. The selenium-mercury paradigm reframes the same data as evidence of alignment. The benefits do not occur despite mercury; they occur because protection is preserved.

Attempts to reconcile these observations within a mercury-centric framework often rely on narrow categorizations—distinguishing "safe" from "unsafe" species, or emphasizing moderation without explanation. While these approaches may mitigate risk at the margins, they fail to address the underlying chemistry that governs outcome. Selenium's absence from the explanatory model leaves guidance perpetually reactive, adjusting boundaries without clarifying mechanism.

Population data also reveal the unintended effects of restrictive messaging. In regions where fish consumption has declined in response to advisories, corresponding changes in nutrient intake have been observed. Selenium and omega-3 fatty acid levels decrease, while reliance on alternative protein sources increases. These shifts do not necessarily reduce exposure to all contaminants, but they do reduce access to nutrients with well-established protective roles. The net effect is often a redistribution of risk rather than its elimination.

The consistency of these patterns across diverse contexts suggests that the issue is not one of isolated anomalies, but of systemic misunderstanding. When a model fails repeatedly to predict observed outcomes, the appropriate response is not to refine its parameters indefinitely, but to question its premises.

Mercury-centric toxicology predicts harm that does not consistently appear. A relational model that integrates selenium predicts resilience that is widely observed. The choice between them is not ideological; it is empirical.

Population-level corroboration does not provide the drama of controlled experiments, but it offers something more durable: convergence. When cultural practice, epidemiology, and biochemical mechanism point in the same direction, the burden of proof shifts. The question becomes not whether selenium matters, but why it has remained peripheral for so long. The answer lies less in the data than in the narratives used to interpret it.

Integrating these population patterns into guidance requires humility. It requires acknowledging that long-standing dietary practices may embody biochemical wisdom that formal models have overlooked. It also requires resisting the impulse to attribute resilience to factors that preserve the existing frame. Selenium's role does not negate the importance of monitoring exposure. It contextualizes it, explaining why exposure alone has proven to be a poor predictor of outcome.

When population data are allowed to speak without distortion, they reinforce a simple conclusion: systems that maintain protective balance outperform those that focus narrowly on avoidance. Fish-consuming populations thrive not because they are immune to mercury, but because their diets sustain the defenses mercury would otherwise undermine. The data do not argue for recklessness. They argue for coherence.

This coherence has been present all along, visible to anyone willing to look beyond isolated variables. The persistence of contradictory outcomes is not a mystery to be solved, but a message to be heard. When theory and observation diverge consistently, it is the theory that must yield. Population evidence has been signaling this for decades. The selenium–mercury paradigm simply listens.

Clinical translation is often where sound theory falters, not because the theory is wrong, but because it is delivered in the wrong form. Models that arrive as mandates provoke resistance. Lists invite misinterpretation. Rules encourage compliance rather than understanding. In environments already saturated with guidance, adding another layer rarely clarifies. What clinicians require is not more instruction, but a framework that reorganizes what they already know into something coherent and usable.

The selenium–mercury paradigm lends itself to this kind of translation precisely because it does not depend on novelty. It does not ask clinicians to abandon established concerns about exposure, nor does it require them to memorize new thresholds or categories. It simply restores a missing variable to the frame and allows existing knowledge to fall back into alignment. Mercury remains relevant. Fish consumption remains variable. Pregnancy remains a period of heightened sensitivity. What changes is the logic that connects these facts.

From a clinical perspective, the most consequential shift is from restriction-based counseling to sufficiency-aware counseling. This does not mean encouraging indiscriminate intake or dismissing individual circumstances. It means recognizing that advising avoidance without addressing what that avoidance removes is incomplete care. When selenium's role is understood, questions about fish consumption can be reframed around protection rather than prohibition. The conversation becomes less about policing exposure and more about ensuring resilience.

This reframing simplifies clinical decision-making rather than complicating it. Instead of navigating complex species lists or debating marginal differences in mercury concentration, clinicians can focus on maintaining nutritional adequacy during periods of high demand. Selenium sufficiency becomes a stabilizing reference point. Where dietary intake supports it, concern about trace exposure diminishes. Where intake is uncertain or restricted, the need for deliberate replacement becomes clear.

Importantly, this approach does not require uniform recommendations. It accommodates variation without sacrificing coherence. Patients with dietary restrictions, allergies, or cultural preferences can still be counseled effectively, provided the protective role of selenium is acknowledged and addressed. The paradigm does not dictate a single diet; it clarifies what any diet must provide to preserve protection.

The shift also reduces the moral burden often felt in clinical encounters around pregnancy. Counseling framed around avoidance can inadvertently amplify anxiety, placing responsibility on patients to navigate risk without adequate context. Sufficiency-aware counseling distributes that responsibility more evenly. It acknowledges uncertainty while emphasizing what is known and controllable. The implicit message shifts from “be careful” to “ensure protection,” a change that alters both tone and outcome.

Clinicians are accustomed to managing conditional risk. They routinely weigh benefits and harms in contexts where certainty is unattainable. The selenium–mercury paradigm aligns with this practice by making conditionality explicit. Mercury’s impact is conditional on selenium availability.

Risk is conditional on balance. Guidance that reflects this reality feels familiar rather than disruptive. It fits within existing clinical reasoning rather than challenging it.

This fit is crucial for adoption. Paradigms do not take hold because they are correct; they take hold because they are usable. A model that requires clinicians to defend it against institutional guidance is unlikely to spread. A model that quietly explains why observed outcomes make sense, and why existing guidance sometimes falls short, integrates itself without confrontation. Selenium's role can be acknowledged in practice long before it is formally codified in policy.

Clinical translation also benefits from restraint. The paradigm does not need to be announced explicitly to be effective. It can inform how questions are asked, how dietary histories are interpreted, and how concerns are contextualized. Over time, this shifts practice organically. Patients receive clearer explanations. Anxiety diminishes. Nutritional adequacy improves. None of this requires new protocols; it requires a corrected frame.

The paradigm also clarifies when caution is warranted. In cases of unusually high exposure or compromised nutritional status, selenium sufficiency may be challenged. Recognizing this allows clinicians to identify genuine risk without resorting to blanket restriction. It sharpens rather than dulls judgment. The absence of selenium from the model, by contrast, blunts judgment by forcing all scenarios into the same restrictive mold.

Translation at the clinical level is ultimately about trust. Patients trust clinicians who can explain risk in a way that feels grounded and proportional. Clinicians trust frameworks that align with observed outcomes and biological plausibility. The selenium–mercury paradigm earns this trust by reconciling chemistry with experience. It explains why many patients thrive despite exposure that guidance deems concerning, and why some struggle when protection is eroded.

This reconciliation has implications for training as well as practice. When relational biology is introduced as a lens rather than an exception, it encourages a more integrated understanding of nutrition and toxicology. Selenium becomes a case study in how protection operates, rather than an isolated fact to be memorized. The lesson extends beyond this specific paradigm, reinforcing a way of thinking that is applicable across domains.

Clinical translation does not demand perfection. It demands coherence. The selenium–mercury paradigm offers that coherence by aligning guidance with the way the body actually manages risk. It reduces the cognitive load on clinicians by replacing fragmented rules with a single, unifying principle: maintain protection where demand is greatest. Everything else follows from that.

When translation succeeds at this level, formal adoption becomes almost unnecessary. Practice evolves, outcomes improve, and the paradigm embeds itself quietly. The chemistry does not change. What changes is the frame through which it is understood. In medicine, that change is often the difference between guidance that is followed and guidance that is merely tolerated.

Public guidance succeeds or fails less on the strength of its evidence than on the stability of the narrative that carries it. Once a message is established, particularly one framed around protection of the vulnerable, it acquires moral weight that resists revision. This is especially true in nutrition and pregnancy, where guidance is absorbed not as provisional instruction but as ethical obligation. In such environments, correction must be handled with restraint. Loud reversals provoke distrust. Quiet clarification, by contrast, allows systems to realign without fracture.

The selenium–mercury paradigm poses a particular challenge for policy because it does not lend itself to the familiar architecture of warnings. It does not identify a single agent to avoid or a simple threshold to observe. It reframes risk as conditional and protection as active, requiring communication that privileges understanding over compliance. This runs counter to the way public guidance has traditionally been constructed, where simplicity is often achieved by omission rather than integration.

Yet simplicity achieved through omission carries its own cost. When guidance focuses narrowly on avoidance, it trains the public to equate safety with subtraction. Over time, this logic becomes reflexive. Foods are categorized as dangerous or safe without reference to context. Nutritional adequacy is assumed to persist unless explicitly threatened. In the case of selenium, this assumption has proven false. The absence of overt deficiency symptoms allowed sufficiency to be taken for granted, even as guidance eroded one of its primary dietary sources.

Policy-level communication rarely acknowledges what guidance removes, focusing instead on what it prevents. This asymmetry is consequential. Advisories that discourage fish consumption do not typically articulate the nutritional tradeoffs involved, nor do they provide a framework for replacing what is lost. The public is left with a sense of risk reduction that may not correspond to biological reality. In pregnancy, where dietary decisions are already fraught, this imbalance amplifies anxiety without delivering clarity.

A policy approach informed by the selenium–mercury paradigm does not require dramatic revision. It requires recalibration. Rather than issuing new prohibitions, guidance can emphasize maintenance of protection alongside exposure awareness. This shift does not weaken precaution; it strengthens it by aligning precaution with function. The message becomes less about fear of contamination and more about sustaining the systems that manage it.

Such recalibration must be subtle to be effective. Institutions are understandably cautious about appearing to retreat from prior warnings. However, reframing guidance around balance rather than avoidance allows for continuity without contradiction. Mercury remains a consideration. Fish consumption remains nuanced. What changes is the explanatory context. Selenium is acknowledged not as an afterthought, but as a central factor in determining outcome.

Public guidance that integrates this context can reduce confusion without inviting backlash. When people understand why certain foods are beneficial despite containing trace contaminants, trust is restored. The guidance feels grounded rather than arbitrary. Importantly, this understanding does not require technical detail. It requires a narrative that reflects relational biology rather than isolated threat.

The challenge lies in communicating conditionality without overwhelming. Public messages are necessarily compressed, but compression need not distort. A focus on protection provides a stable anchor. When guidance emphasizes that certain nutrients preserve the body's defenses—particularly during pregnancy—it creates a positive orientation that can accommodate nuance. Avoidance becomes one tool among many, not the defining principle.

Policy inertia often stems from fear of misinterpretation. Officials worry that acknowledging protective mechanisms will be taken as license for excess. This concern reflects a lack of trust in the public's capacity to understand balance. Yet experience suggests that clarity reduces misuse more effectively than prohibition. When people are given a coherent explanation, they are better equipped to make proportionate choices. When they are given only warnings, they are left to navigate tradeoffs without guidance.

The selenium–mercury paradigm also offers a way to harmonize guidance across domains that have drifted apart. Nutrition, toxicology, and maternal health often issue advice independently, leading to fragmentation. Integrating protection as a unifying concept allows these domains to speak a common language. Selenium's role bridges them, connecting dietary patterns with biochemical defense and developmental outcomes. Policy that reflects this integration feels less like a patchwork and more like a system.

Quiet correction is particularly important in environments where trust has been strained by shifting recommendations. Rather than announcing paradigm shifts, institutions can allow updated frameworks to permeate training materials, clinician education, and public-facing resources gradually. Over time, the emphasis shifts without the need for formal repudiation of earlier guidance. The story evolves, and with it, behavior.

This approach respects the reality that policy operates within social as well as scientific constraints. Evidence alone does not drive change; narratives do. By recasting the narrative around protection, the selenium-mercury paradigm aligns with deeply held values—care for the vulnerable, responsibility during pregnancy, stewardship of health—without triggering defensiveness. It does not accuse. It clarifies.

The effectiveness of such guidance can be measured not by compliance rates, but by outcomes. When nutritional adequacy improves, anxiety diminishes, and developmental markers align with expectation, the guidance has done its work. These improvements may not be immediately attributed to a paradigm shift, but attribution is not the goal. Alignment is.

Policy informed by relational biology accepts that risk cannot be eliminated, only managed. It recognizes that protection must be sustained, not assumed. In this context, selenium is not a special case, but an exemplar—a reminder that the body's defenses are active participants in health, not passive recipients of exposure. Guidance that acknowledges this participation becomes more resilient, capable of adapting as evidence evolves without destabilizing trust.

Ultimately, the success of public guidance depends on its ability to reflect reality without provoking fear. The selenium–mercury paradigm offers a path to that balance. By emphasizing protection alongside awareness, it allows institutions to fulfill their mandate without oversimplifying the chemistry that governs outcome. The correction need not be loud. It need only be accurate.

Restoration rarely arrives as resolution. It does not tie arguments neatly or declare winners. Instead, it reestablishes proportion. What was overstated recedes. What was neglected returns to view. The system, once distorted by emphasis, resumes its ordinary work. This is the nature of protection when it is finally understood—not as intervention, but as alignment.

The selenium–mercury paradigm does not culminate in a directive. It culminates in a recognition. The chemistry that governs risk and resilience has been consistent throughout, indifferent to the narratives imposed upon it. Selenium has continued to support enzymatic defense. Mercury has continued to bind competitively. Development has continued to demand protection on a schedule that does not wait for guidance to catch up. The failure was never biochemical. It was interpretive.

What restoration offers is not absolution for past guidance, but relief from its constraints. Once protection is restored to the center of the frame, the urgency that once animated avoidance loses its force. Mercury no longer needs to be banished from consideration because it no longer carries explanatory weight disproportionate to its role. Fish no longer need to be defended as exceptions because their coherence is visible. Pregnancy no longer needs to be navigated as a field of hidden threats, but as a period with clear, if unforgiving, requirements.

This shift has a calming effect precisely because it removes spectacle. There is no dramatic reversal to announce, no reckoning to stage. The paradigm corrects by subsuming, not by overturning. Everything that was previously true remains true, but no longer sufficient. Exposure matters, but only in relation to protection. Caution matters, but only when it preserves adequacy. Guidance matters, but only when it reflects the system it seeks to govern.

The temptation to conclude with instruction is strong, particularly in a subject so closely tied to care. Yet instruction is unnecessary when understanding is complete. The body does not require reminders to protect itself. It requires the conditions that allow protection to occur. Selenium provides one of those conditions. When it is present in sufficient quantity, enzymatic systems function as intended, oxidative stress is managed, and development proceeds without accumulating silent compromise. When it is absent, no amount of warning can compensate.

Restoration also clarifies why the problem persisted despite ample evidence. The paradigm was not missing because it was unknown. It was missing because it lacked drama. It offered no villain, no threshold, no call to action. It asked only that balance be maintained. In environments attuned to crisis, such a request struggles to compete. Yet balance, once restored, resolves what crisis narratives cannot. It absorbs uncertainty rather than amplifying it.

There is a humility inherent in this outcome. The solution does not originate in policy innovation or technological intervention. It originates in chemistry already at work. Selenium's role did not need to be invented or engineered. It needed to be recognized. This recognition does not elevate selenium beyond its station; it returns it to it. Protection is neither heroic nor optional. It is infrastructural.

This infrastructural quality is what makes restoration durable. Systems built around avoidance must continually adjust as new threats are identified. Systems built around protection remain stable because they address vulnerability at its source. The selenium-mercury paradigm belongs to the latter category. It does not chase contaminants. It sustains defenses. In doing so, it renders many concerns manageable without fanfare.

The implications extend beyond the specifics of fish consumption or pregnancy guidance. They speak to a broader tendency in modern health discourse to privilege what can be measured loudly over what operates quietly. Restoration challenges this tendency by insisting that effectiveness does not require visibility. The most consequential protections are often those that leave no trace precisely because they prevent disruption from arising.

In this light, the earlier misalignment appears less as a failure of knowledge than as a failure of emphasis. Mercury was emphasized because it was conspicuous. Selenium was neglected because it was not. The correction does not require new discoveries. It requires a redistribution of attention. Once attention is properly allocated, the system resolves itself.

This resolution is not an endpoint. It does not close inquiry or foreclose refinement. It simply establishes a frame within which future questions can be asked more productively. Exposure can be evaluated without distortion. Dietary patterns can be assessed without fear. Developmental outcomes can be interpreted without resorting to speculation. The paradigm does not answer every question. It ensures that the right questions are asked.

Restoration also carries a quiet ethical dimension. When guidance aligns with biology, responsibility is shared rather than displaced. Individuals are not burdened with navigating contradictions. Clinicians are not forced to reconcile incompatible models. Institutions are not compelled to defend simplifications they know to be incomplete. The system becomes more humane by becoming more accurate.

In pregnancy, where the stakes are highest and the margin for error is smallest, this humanity matters. The developing brain does not benefit from anxiety layered atop insufficiency. It benefits from continuity—of nutrients, of protection, of care grounded in reality rather than apprehension. Selenium's role in providing that continuity is unremarkable in its mechanics and profound in its consequences. It exemplifies how protection operates when allowed to do so.

The restoration offered by this paradigm is therefore quiet by necessity. Loud conclusions would betray its nature. Protection does not announce itself when it succeeds. It simply leaves the system intact. The absence of harm is its signature. The presence of resilience is its proof.

What remains, then, is not a mandate but a frame. A way of seeing that accommodates complexity without amplifying fear. A way of guiding that preserves protection rather than eroding it. A way of understanding that allows chemistry to explain what narrative obscured.

The selenium-mercury paradigm resolves the problem not by arguing it away, but by placing it where it belongs—within a system capable of managing it. Once placed there, the problem diminishes, not because it was trivial, but because it was finally understood. Protection, restored to its rightful place, does what it has always done best. It works quietly, completely, and without spectacle.

Acknowledgments

The ideas explored here are indebted to the work of researchers in toxicology, nutrition, and developmental biology whose careful studies have long pointed toward relational explanations for outcomes often attributed to isolated causes. Particular appreciation is extended to those whose work on selenium-dependent enzymatic systems helped clarify the chemistry underlying this paradigm.

Any errors of interpretation remain the responsibility of the author.

About the Author

Joe Cozart is the founder of GMJoe™ Consulting, where his work focuses on clarity in complex systems across science, policy, and strategic communication. His writing explores the intersection of chemistry, development, and institutional decision-making, with particular attention to how incomplete frames persist in public guidance.

He lives and works in the United States.

Closing Note

The absence of harm is rarely noticed. Yet it is often the most reliable indicator that protection is functioning as intended. This work is offered in that spirit.

Made in the USA
Monee, IL
04 November 2026

67387321R10036